Much of the credit for this book must go to my friend Dave Gingery, whose books on the Charcoal Foundry and the Build Your Own Metal Shop From Scrap series gave me the confidence and where-with-all to build the foundry and create some wonderful shop tools, and to my wife, Patsy, who provided the ideas and advice for creation of many of the ornamental projects covered in this book.

D0963948

WARNING

Remember that the materials and methods described here are from another era. Workers were less safety conscious then, and some methods may be downright dangerous. Be careful! Use good solid judgement in your work, and think ahead. Lindsay Publications, Inc. has not tested these methods and materials and does not endorse them. Our job is merely to pass along to you information from another era. Safety is your responsibility.

Write for a catalog or other unusual books available from:

Lindsay Publications, Inc.
PO Box 12
Bradley, IL 60915-0012

ORNAMENTAL

METAL
CASTING

R.E. WHITMOYER

Lindsay Publications Inc.

Ornamental Metal Casting

by Robert E. Whitmoyer

Copyright © 1986 by Lindsay Publications, Inc., Bradley, IL 60915

ISBN 0-917014-43-0

5 6 7 8 9 0

Table of Contents

Introduction

Since the very beginning when man learned to cast metal, he has decorated even simplest objects with ornamentation to make them pleasing to the eye as well as functional. Realizing that metal was both durable and beautiful, man began casting sculptures and adding decoration to even the most mundane metal castings.

Many of us at one time or another have wanted to create metal objects but did not have the techniques, equipment or knowledge to do so. In addition, many people feel that working metal is too difficult or expensive for the home craftsman. But that is just not the case. Whether it is a simple wall plaque, or cast metal chess set or a full-size working fountain for your back yard, you can create these prized ornamental works of art. Using your own patterns and ideas you can create exactly the objects you want at a small fraction of their commercial price if they're available at all.

Anyone can master the simple techniques outlined in this book and use them to achieve professional results. Don't be afraid of the process. Take each step, one at a time, and learn as you go. You'll find several supplemental books listed in the index should you wish to further develop your skills.

Take your time in following the steps in this manual. Don't hurry. Before you realize it you will be able to produce almost any type of casting you wish. You will be able to create nearly anything you want. The most pleasant part is that your pocketbook will have hardly noticed the effort.

The book will be an excellent guide for those who wish to make some castings for the home, for shop teachers wishing to demonstrate early foundry methods, for school art departments, and others who have had trouble getting castings poured at commercial foundries.

The author is neither a professional foundryman or engineer and assumes no responsibility for the safety of the procedures outlined in this book. Foundry work is, by nature, very dangerous, but so is a pot boiling water. Use extreme care in your work.

Part One
The Foundry

Chapter 1
Melting Methods

In order to cast metal we must melt it and have a mold ready into which to pour it.

First, let's consider the methods of melting metal. There are a number of commercial furnaces available which will melt aluminum or brass but these are relatively expensive, certainly in excess of several hundred dollars. If you can afford this type of equipment, by all means buy it. But the furnace to be described is for those of us on a limited budget. There are economical alternatives for melting pewter, potmetal or lead, which are the most popular of the low melting temperature metals. (Gold, silver and platinum are not discussed, although the method mentioned below will handle these as quickly as they do pewter.)

Low temperature alloys and metals can be melted using an ordinary gas ring. Gas burners from kitchen stoves are not recommended due to their lightweight construction. The single burner gas hot plate of the old cast iron variety is an excellent choice for this operation, and these can be found in junk shops nearly everywhere. (Fig. 1-1) This will allow you to place the burner in a sand bed where an accidental spill will not be a disaster as it would be in the family kitchen. This gas hot plate can be hooked into a small (camping trailer size) propane tank. Parts from a junk gas barbe-

FIGURE 1-1 Use ordinary gas ring to melt low temperature alloys and metals. The single burner gas hot plate is an excellent choice.

que can be used to hook up the tank and to re-jet the hot plate, if necessary.

I was able to obtain an excellent melting ladle with wooden handles and a pig mold at a local auction for $10 (Fig. 1-1). If you cannot find something similar, a good ladle can be fabricated by welding a bottom onto a 6'' length of 4-6'' diameter thin-wall steel pipe and by welding a 12-16'' handle (3/4'' steel pipe) on one side of the pipe. This ladle can be further improved by heating the top of the pipe with a torch and forming a pouring lip in the far side (not in the front as you are holding the handle).

Having obtained a gas ring, made a melting ladle, and scrounged up an old cupcake pan and a handful of junk wheel weights (lead) from your local gas station, you are now ready to try your furnace.

Pick an area outside, away from buildings and shrubbery. Clear all flammable material from the immediate area and set up your equipment. Check your set-up again, and then light your gas ring and place the ladle on the burner. Drop several of the wheel weights into the ladle, and then wait for about 5 to 20 minutes for the lead to melt. As the lead melts a lot of junk will float to the top. This is called dross or slag. Skim this off the top of the melted metal with a bent tablespoon and place the hot slag in one of the cupcake spaces in the pan.

Grasp the ladle with both hands while WEARING HEAVY HEAT-RESISTANT GLOVES. Lift the ladle off the burner, and carefully pour the melted metal into the cupcake pan, filling each depression about 3/4 full. Place the ladle back on the gas burner, and remove your gloves. Turn off the gas burner, and shut off the propane tank.

By casting scrap into pigs for later use, we are able to clean up the metal considerably which gives us better castings in the end. You can collect, melt and mold a quantity of lead, pewter and potmetal into pigs. Practice with your melting equipment to get a feel for the time of the melt, the quantity that your ladle holds, and at the same time, to build up a quantity of good clean metal for melting when we begin to pour useful castings. In general, I have found that this process of melting, producing pigs, and re-melting eliminates the need for fluxes and degassing agents which are used extensively in commercial foundries but are often hard to use in the home foundry.

Where can the home foundryman find metal to melt? You can try to buy commercial pigs of metal from a foundry. These are often expensive if available at all.

Assuming the operators are willing to sell, you can get a wide variety of suitable metal from your friendly neighborhood junkyard. It's the best department store ever invented for the home hobbiest.

LEAD: Found in wheel weights, old plumbing pipe, boat anchors, etc. It is gray, very heavy and scrapes easily with a pocket knife to a bright silver finish.

PEWTER: Pewter is a combination of lead, tin and antimony. It was used extensively before 1900 in many kitchen items and other metal objects. It is currently used only for small art works.

POTMETAL: Found in decorative metal plaques, small appliance cases, etc. It is very light and breaks easily with a hammer.

ALUMINUM: Found in auto pistons, lawn chairs, storm doors, etc.

BRASS: Found in old plumbing valves, scrap copper pipe, etc.

Chapter 2
Furnace and Crucible
Construction and Use

FIGURE 2-1 Furnace built from a standard metal garbage can by the author. It has a 2½ quart capacity.

CRUCIBLE FURNACE

The simplest and cheapest way to melt and pour aluminum or brass is with a crucible furnace. The metal is held in a steel or ceramic crucible which is

7

placed inside the furnace during melting. There are suppliers who will sell you one of these furnaces, such as Pyramid Products Co., Phoenix, AZ, who supplies small operators and hobbiests.

David Gingery has written a book (refer to his series of books, "Build Your Own Metal Working Shop from Scrap," from Lindsay Publications, Inc., Bradley, IL) called the "Charcoal Foundry" in which he describes in detail the construction of a one quart capacity furnace using a five gallon metal can.

I found that some ornamental castings exceed the one quart capacity. At times there are large numbers of castings to pour (the fountain requires 23 separate castings) making greater capacity desirable. The furnace that I built has a 2-1/2 quart capacity and is as large as can be safely manipulated by one determined man. (Fig. 2-1).

BUILDING THE FURNACE

The body of the furnace consists of a standard metal garbage can 23" high and 17-1/2" diameter at the top with a 2-1/2" thick refractory lining on the inside and tweer (traditionally spelled "tuyere") hole near the bottom for the air blast. The top is made of the same refractory lining.

There are a couple of things that should be built before the furnace itself.

First, we need to build the air blast unit which consists of three parts: (1) a small squirrel cage fan, not more than 40 CFM, (2) a three foot piece of metal downspouting, and (3) a spouting connector fitting.

A spouting connector is a fitting that connects a vertical downspout to the horizontal gutter that runs along your roof. It will be used to connect and remove the air blast from the furnace with ease.

Set the connector flange on the outside of the garbage can about 4" from the bottom lip to the bottom of the connector. Using a 1/4" drill, drill a closely spaced series of holes around this circle. Break out the metal inside of the circle to form the air blast or tweer hole. Pop rivet or bolt the connector flange into place. Next, cut a 3'' diameter piece of wood about 6'' long and tapered toward one end so that it fits thru the connector flange and into the garbage can about 3''. This will be the tweer hole form around which the refractory mix will flow when rammed into the can.

Now drill 6 to 8 evenly spaced 1/4'' holes through the sides of the garbage can, one circle of 8 about 6'' from the top and and another circle of 8 about 12'' up from the bottom. Insert a 1/4-20 x 1-1/2'' bolt in each hole and tighten with a nut on the inside of the can. The bolt should stick into the can about 1-1/4'' to help hold the refractory in place after it has been fired.

Also at this point, it would be a good idea to drill 1/4'' holes through each side of the handle brackets of the garbage can. Use 1/4-20 bolts 1/2'' long through these holes to reinforce the mounting rivets or welds. Spot welds tend to come loose later on. (Fig. 2-2)

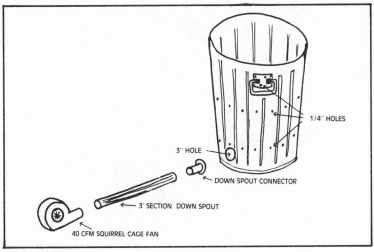

1/4'' HOLES

3'' HOLE

DOWN SPOUT CONNECTOR

3' SECTION DOWN SPOUT

40 CFM SQUIRREL CAGE FAN

FIGURE 2-2 Air blast unit.

Next, we need a form for the furnace lid. The form consists of a long strip of heavy gauge sheet metal 3''

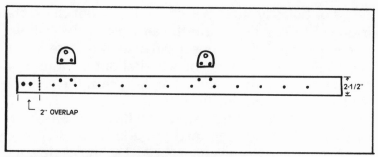

FIGURE 2-3 Heavy gauge sheet metal lid hoop.

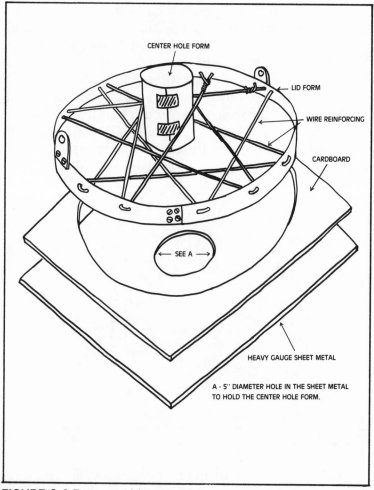

FIGURE 2-4 Furnace Lid assembly.

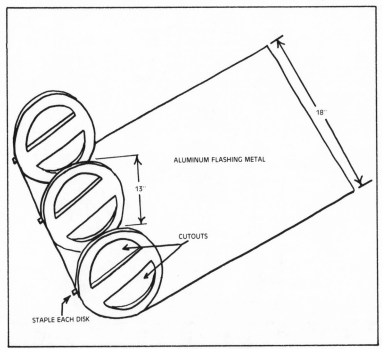

FIGURE 2-5 Inside form for furnace assembly.

wide and long enough so that when it is bent into a circle with the ends overlapped about 3″, it will fit into the top of the garbage can with 1″ clearance on each side. Drill the holes for the reinforcing wire and handle straps. (Fig. 2-3) The handle is made from a 1/8″ steel brazing rod. Using small bolts or heavy aluminum pop rivets and some old clothes hanger wire, assemble the form as shown in Fig. 2-4.

The lid form is assembled complete with handles. Cut a circle the same diameter as the lid form in a large sheet of heavy cardboard. Slip the lid form over it. The cardboard will keep the form from being bent out of round during the ramming of the refractory. Place entire assembly on a piece of heavy gauge sheet metal (1/8″ thick) with a 5″ hole in the center.

The 5″ aluminum center hole form is made by creating a circle of aluminum flashing metal 4″ high. Tape it along the seam. Besides providing a vent hole

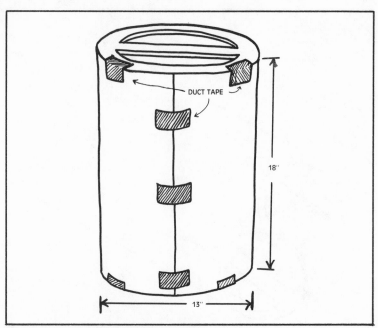

FIGURE 2-6 Finished inside furnace form.

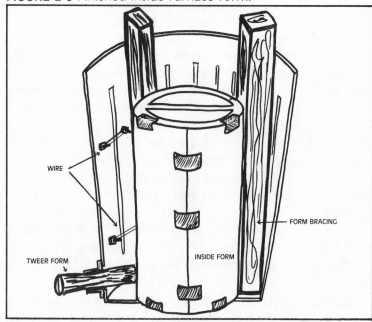

FIGURE 2-7 Positioning of inside furnace form. Ready for refractory material.

in the refractory, this form also provides stability to the whole lid form. (Fig. 2-4)

The last item is the inside form of the furnace. This is made up of three 13" diameter 1/2" thick plywood discs and a piece of aluminum flashing metal of the type used around chimneys and gutters of houses. Cut a strip of 18" wide flashing metal long enough to reach around the plywood discs with a 2" overlap. Assemble the form as shown in Fig. 2-5 and detailed in the steps below.

1. Lay a sheet of flashing metal on the floor, and fasten each disc to the end with a staple gun.

2. Roll the aluminum flashing tightly around the discs, and tape the ends with duct tape as shown.

3. Finish taping as shown in Fig. 2-6, and set the finished form to one side for now.

Do not overdo the taping because the discs will have to be removed from the metal after the lining is in place in the furnace. Set the form in the bottom of the garbage can, and cut four piece of wood, 20" long and just wide enough to slip them in between the form and the sides of the can. This will hold the form in place during the ramming of the refractory. (Fig. 2-7)

REFRACTORY MATERIAL

This is going to be a two day job so perhaps starting on a Friday evening would be a good idea. The refractory material can be purchased from local supply houses in your area. Dave Gingery has found a type called Kaiser's IRC castable refractory to be a very satisfactory product. I could not find this in my area and the available materials were quite expensive so we made our refractory on site, out of cheap, available materials. The refractory consists of clean silica sand and fire clay mixed two parts sand to one part fire clay ratio.

You'll need about two gallons of finely crushed fire brick or ceramic flue liners. I went to one of the local masonry supply houses, and they gave me several broken flue liners. I broke the large pieces up into dime sized pieces and mixed them into the refractory mix. This is called "grog" and it results in a stronger lining. Less cracking will occur when the lining is fired during the later stages of furnace construction. The following is a step-by-step description for ramming up and firing the furnace.

1. Spread a sheet of plastic about 8 ft. square on a flat dry surface like your garage floor, and tape the corners down with a little duct tape.

2. Pour about 10 gallons of sand onto the center of the plastic, and spread it out about 3'' thick.

3. Carefully add about 5 gallons of fire clay, and mix in carefully so as not to lose any more of the clay than necessary. The clay becomes airborne with ease. Next, mix in the grog (crushed flue liners).

4. Using a garden sprinkling can, add water in small amounts, mixing well after each addition until the whole mixture has the consistency of very stiff mortar.

5. Fold the edges of the plastic over the pile. Let it set until morning.

6. Check your can and lid form to make sure that it is complete. Reread this chapter to make sure that all necessary preparations for ramming up the furnace have been completed. Say a small prayer for a sunny day tomorrow, and relax until morning.

RAMMING UP THE FURNACE

Set the garbage can in an open area outside, away from buildings and anything flammable. Move the lid form, bracing, etc. nearby where it will be handy. Set the form in place inside the garbage can, and put the bracing in place. Set the tapered plug in place through the tweer hole. (Fig. 2-7)

Scoop up a bucket full of the refractory material, and using a small mason's trowel, put refractory mix into the space between the can sides and the aluminum form to a depth of about 4". Using a 3/4" wooden dowel, tamp the mix down firmly. This is called "ramming." Be careful not to dent or move the inside form.

Move the braces up, work refractory in, and ram it in place. Then move the braces up, and fill in the holes. Repeat this until the refractory is level with the top of the form. Trowel the top smooth, and let it set while you ram up the lid.

RAMMING THE LID

Using the same general method used to ram up the furnace walls, ram up the lid. Be careful not to bend the wire or distort the center form, but be sure to get all the bubbles and voids worked out of the wet refractory. Trowel the top of the lid smooth. Next, take a sharp utility knife, and slit two sides of the cardboard lid form. Remove it carefully. If you leave the cardboard lid form in place it may catch fire when the lid is cured.

Now go back to the furnace, and carefully remove the wood discs from the form. Peel the flashing metal away from the refractory, which should stay in place. Smooth the sides of the refractory GENTLY with the trowel, if needed.

Next, trowel refractory onto the bottom of the furnace until it comes up to a level about 1/4" below the tweer hole plug. Carefully remove the wooden plug from the tweer hole, and smooth the inside of the hole with your fingers.

FIRING THE LINING

Place some paper and wood chips in the bottom of the furnace. Add a little charcoal lighter, and set the whole mess on fire. When it is burning well, start adding more and bigger pieces of wood (pieces no larger than your hand) to the fire. The idea is to keep adding wood until the entire furnace is filled to the top with burning wood. A lot of steam will come out of the refractory mix but this is normal.

When the furnace is full of burning wood, place the lid on top of the furnace. Continue to add wood to the fire through the hole in the lid for about an hour or until the lid stop steaming.

When the lid has stopped steaming, remove it from the furnace, protecting your hands with heavy gloves. Set the lid on several fire bricks. Add about 2" of charcoal briquets (the same kind you use for cooking steaks on the grill) to the coals in the bottom of the furnace. These should light quickly.

Once the charcoal is burning well, fill the furnace to the top of the lining with charcoal. Carry the sheet metal on which the lid form lies to the side of the furnace. Carefully slide the lid onto the top of the furnace. Remember to remove the inside form from the center hole. Fold the handle down to the side to keep it from getting too hot. Connect the air fan to the tweer hole, and turn on the air blast.

In about 15 minutes, all the charcoal will be burning. The gases coming out of the hole in the lid will be in excess of 3000° F. so be very careful. From here on out, there will be plenty of opportunities to burn yourself with your foundry. Be very careful.

When the charcoal has burned itself out, turn off the air blast, and let the furnace cool down on its own for about 4 hours.

Once cool, remove the furnace lid, and set it aside. Dump the ashes by turning the furnace upside down.

Now you can set the furnace upright and inspect the lining for cracks. There may be some fine cracks, but probably no major ones. If any major cracks have appeared in the lining or the lid, wet the area of the crack with some water and patch it with the leftover refractory. Refire the furnace, using only charcoal this time. Firing with wood is not necessary after the first time.

Your furnace is now ready for use. I've used mine for four years without any repair at all, although I confess that it could use a bit of patching around the air hole. It's still in good shape, and you can expect that yours will give the same service.

CRUCIBLE CONSTRUCTION

The simplest and safest crucible I have found for this size furnace is a scaled down version of a steel mill crane ladle. It is constructed as follows:

1. Obtain a 9" piece of 6" outside diameter, 1/4" wall steel pipe, and weld a 1/4" plate on the bottom. Weld inside and out. This weld must be tight, so if you're not sure of yourself, have a professional do it.

2. Drill three 1/2" holes in each side as shown in Fig. 1-10.

3. Using a hacksaw, cut out the metal between the holes as shown in Fig. 2-8.

4. Heat the front side with a torch until bright red, and using a ballpeen hammer, forge a pouring spout or lip on the crucible. (See Fig. 2-8)

5. Bend a 3" piece of 1/2" steel rod into a "U" shape, and weld it to the bottom rear of the crucible. (Fig. 2-8)

6. Next, obtain a 6-1/2 ft. piece of 3/8" steel rod, and cut it into the following: one 21" piece, one 8" piece, and one 45" piece.

7. Heat and bend the 21" section into a hook and handle as shown in Fig. 2-9.

8. Next bend the 45" section into a long "U" shape with at least 8" of flat handle at the top. Weld the 8" section across the open end of the "U". (Fig. 2-9) This must be a good weld, chip the slag off, and inspect it well.

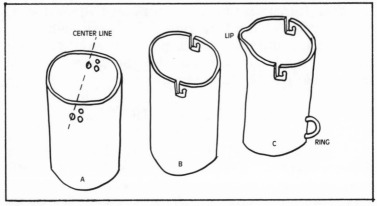

FIGURE 2-8 Crucible is scaled-down version of a steel mill crane ladle.

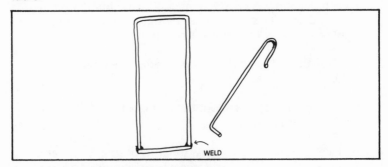

FIGURE 2-9 Crucible handling rods.

The crucible is now ready to be used. It is used as shown in Fig. 2-10. This design allows very good control of the crucible and allows you to pour "away"

18

from you rather than towards you as you would using a traditional crucible and tongs.

You are now almost ready to melt some metal and make a pour. First, we must get some molding sand or a pig mold of some sort into which to pour the molten metal. NEVER LET METAL HARDEN OR FREEZE IN THE POT. Expansion of the metal as it freezes will ruin the crucible.

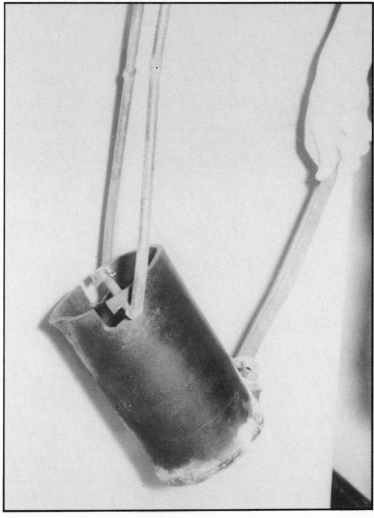

FIGURE 2-10 Crucible ready for use.

Chapter 3
Molding Sand, Flasks, and Odd-Shaped Tools

TOOLS

Every trade has its special tools and equipment. The foundry is no exception. Foundry tools usually seem bizarre to everyone except a "sandcrab." (Sandcrab: traditional name for a foundry molder, or a grown man who still plays in his own sandbox!). While you are looking for some commercial molding sand, free I hope, you can collect the following items:

Wooden potato masher
Small wooden rolling pin
Teaspoon
Old butter knife
Tablespoon
Some hat pins
Long-handled steel spoon or ladle
Two sizes of stiff wire (12" long)
Disposable 50 cc syringe (see your local vet)
Small and large mason's trowel
2 pieces of pipe, 8" long x 1" diameter
2 pieces of pipe, 8" long x 1-1/2" diameter
Box of face masks like those used by autobody painters

Bag of commercial parting compound or swimming pool
  filter material (diatomatous earth)
Selection of wooden dowel rods about 10'' long in a variety of
  diameters up to about 1-1/2''
Small hand bellows, often sold as a fireplace accessory
Couple of old wool socks
Tin cup or old soup can
Pack of cup hooks from the local hardware store to use as
  draw pegs
Old heavy cast type open end wrench for use as a rapper
4 to 6 used fire bricks (like the ones used in fireplaces

If, by now, you have failed to find any commercial
molding sand or are unable to work with the sand you
found (several types of commercial molding sand require power presses to make the molds) you can make
your own. Here is a method for making molding sand
from local materials.

First, obtain about 200 lbs. of the very finest grade
(small grain size) of washed silica sand that you can,
and then filter this through fine window screen to
remove the larger grains and lumps. Purchase an 80
lb. bag of fire clay—same stuff you used for the furnace. Before mixing the clay and sand, you will need
some place to put your molding sand when it's mixed
and tempered (dampened to make it hold its shape).

At this point you should build a molding table
which will hold the sand and provide a work surface
on which to make your molds. Now is the time to
DECIDE EXACTLY WHERE YOUR FOUNDRY IS
GOING TO BE. Once the molding table and sand are
all together as a unit, they become too heavy to move
without a lot of effort.

THE MOLDING TABLE

I made my molding table by constructing a 3' x 5'
box out of 2'' x 6'' plank and nailing a piece of 3/4''

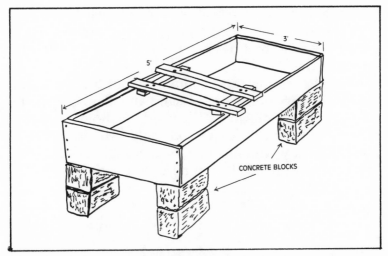

FIGURE 3-1 Molding table and sliding platform.

plywood on the bottom. Then I set the whole thing on concrete blocks high enough to make a comfortable work height. You should construct a sliding platform from 2x4s as shown in Fig. 3-1. Make it so that it will slide along the molding table. I installed the whole thing in a 10' x 10' metal shed purchased from a local discount store. It nicely holds all of my foundry equipment with at least a third of the shed left over for "non-essentials" like lawn mowers, rakes and other yard equipment.

Another device you will need to help in mixing your molding sand is a "riddle," which is nothing more than a coarse sieve or screen. Build a 12" square wooden box with 3" high sides but without top or bottom. Nail a 14" square piece of "hardware cloth" across the bottom. Hardware cloth is the trade name for heavy wire screen with 1/4" square holes, available at hardware stores. Fold the screen up the sides, and staple it in place. See Fig. 3-2. It would be a good idea to make a second riddle with window screen so that you have a fine sieve also.

The most critical part of this whole art is the creation of the molding sand. The mixing procedure is similar to that of the refractory mix. Place a large sheet of plastic on a level surface, and spread out about 100 lbs. of the sand that has been sieved through window screen.

Good molding sand contains about 5 to 15% clay by weight depending on the sand and clay. Clay gives the sand the "stickiness" it needs to hold its shape when damp and make a good mold. Mix in about 10 lbs. of fire clay, and mix in very thoroughly. Mix slowly to prevent the clay from becoming airborne.

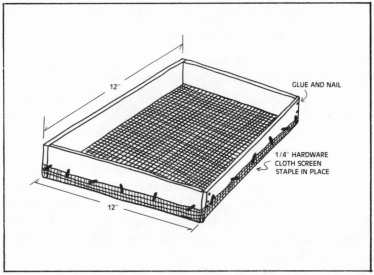

FIGURE 3-2 The "riddle."

Next, using a sprinkling can, add about two quarts of water mixing the pile as you dampen it. This amount should dampen the entire pile of sand, but don't add so much water that the sand begins to feel "wet." If the molding sand is dry in some areas and has large lumps, you can safely add almost another quart of water. At this point the sand should begin to feel damp. Work the sand into a neat pile, and cover with the edges of the plastic. Give yourself an hour break. You'll need the time to explain to your family why you're doing this anyhow!

Next, uncover the pile, and grab a handful of the sand. Squeeze tightly in your hand. One of three things will happen:

A. The sand will all squish out between your fingers. When you open your hand nothing will be left except a layer of sticky sand all over your hand. The sand is too wet. Uncover it, and let it dry overnight. Spread the pile out for better drying.

B. The mixture will crumble leaving you with a handfull of loose sand when you open your hand. In this case, the sand is too dry. Add more water, and try again in an hour.

C. When you open your hand, the sand has formed a compact lump that clearly shows the imprint of your hand and fingers. You will be able to gently break the lump into two pieces without crumbling it, and the break will be clean. SUCCESS! Now we are nearly there...

When the squeeze test nearly matches "C" above, scoot the sand pile to one side of the mixing sheet. Take the riddle that you have constructed, and riddle (seive) a 1/4" layer of the molding sand onto the mixing sheet. Next, using a window screen riddle or a flour sifter like the one your wife has in the kitchen (hint! hint!) sift a 1/8" layer of fire clay on top of the sand. Riddle another 1/4" layer of molding sand onto the mixture. Next, sift a fine coating of wheat flour onto the mixture. Repeat the procedure: Molding sand - fire clay - molding sand - wheat flour - molding sand -fire clay - etc., etc., until you have used up the original pile of sand completely. Then mix the entire pile well using a trowel and the riddle. On the last mixing cycle, retemper (add water) until the sand has that damp feel again. Use the squeeze test, and cover with plastic overnight. The following morning, repeat the squeeze test. The results should be even better than before.

Chapter 4
Molding Your First Casting

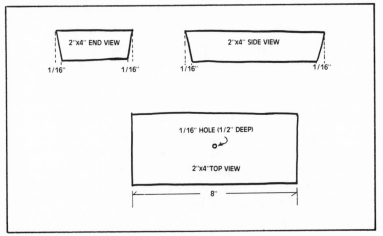

FIGURE 4-1 Scrap 2x4 tapered to fit inside the riddle to serve as pattern to test sand.

LET'S TRY IT ALL OUT

Cut a couple of pieces of 1/4" plywood that will just fit inside the riddles. Place these panels in the bottoms of the riddles.

Cut a scrap piece of 2x4 that will fit inside the sieve boxes with about 2" clearance on each end. We'll use this as a pattern with which to test the sand. With a wood rasp or sanding disc, taper all of the sides of the

2x4 scrap slightly. See Fig. 1-14. Patternmakers call this taper "draft." It allows the pattern to be easily removed from the sand.

Drill a 1/16" hold about 1/2" into the center of the top of the pattern. Take a riddle with the plywood insert in place, and ram in about an inch and a half of molding sand. Place the pattern in the seive box, and ram molding sand firmly all around the pattern until level with the top. Use the potato masher as a ramming tool. Smooth with the trowel, and dampen the sand around the edge of the pattern with water, using the syringe.

The pattern must now be loosened from the sand so that it can be removed. This is called "rapping and drawing." First, screw one of the cup hooks into the hole in the center of the pattern. Using an old wrench or similar tool, tap the draw peg horizontally several times in each direction. Note that this compacts the sand a bit around all the sides of the pattern and the pattern becomes "loose" in the sand. Now grasp the draw peg, and lift the pattern straight up and out of the sand. This is called "drawing the pattern." You should find a rectangular cavity with good clean edges and a smooth bottom. Repeat using the other riddle. We will use these as molds to receive the molten metal from our first melt and run.

FIRING THE FURNACE

It's now time to fire up the furnace. Foundry work using this type of furnace is ALWAYS AN OUTSIDE ACTIVITY due to the large amounts of carbon monoxide given off by burning charcoal.

---

1. Lay a 7" piece of 2" angle iron in the bottom of the furnace to provide an air shaft from the tweer hole to the rear of the furnace. Several 1/2" holes can be drilled to further improve air distribution. (Fig. 4-2)

---

2. Lay in some charcoal, about 2 briquets deep, and light it. Always be careful with the charcoal lighter, and remove it from the area immediately after use.

3. Once the charcoal is burning well, add more. Add enough so that when the crucible is placed in the furnace, the top edge of the crucible is level with the top of the furnace lining.

4. With the crucible in the furnace, pack charcoal around it until the charcoal is level with the top of the crucible.
5. Fill the crucible with broken-up aluminum scrap. Do not pack the scrap in tightly.

6. Put the furnace lid in place, and connect the air blast. Turn on the blower.

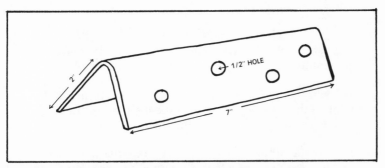

FIGURE 4-2 Angle iron serves as air shaft in furnace.

In about 10 to 30 minutes the crucible will be red hot, and the aluminum will start to melt and form a puddle in the bottom of the crucible. Continue to add scrap aluminum until the crucible is about half full. Using a 3/16'' diameter steel brazing rod about 3 feet long and heavy welders gloves, stir the melt from time to time to hasten the melting. I believe the safest way to do this is to:

1. Turn off the air blast.

2. Remove the furnace top, and set it aside on the fire brick.

29

3. Stir the melt, and add more scrap.

4. Replace the top, and start the air blast.

5. Repeat as needed until the crucible is half full.

6. If the crucible settles, raise it with the lift handle, and add more charcoal to the furnace.

When the crucible is half full of melted aluminum, allow the furnace to run about two minutes, and then turn off the air blast. Remove the blower unit from the furnace.

1. Remove the lid and place it on the fire bricks by the side of the furnace.

2. Stir the melt, and check its consistency. It should stir easily, and the stir rod should have only a thin coating of aluminum or nothing on it when it is withdrawn.

3. Using a large long-handled spoon bent to form a ladle, skim off the slag from the top of the melt. Be careful! This slag is just as hot as the molten metal.

4. Take the lift handle, and insert it into the slots on the crucible. Make certain that it's firmly in place. Hold the lift handle in your left, and pick up the crucible hook with your right hand. (Reverse this if you're left-handed.)

5. Lift the crucible straight up out of the furnace. As the crucible clears the furnace, insert the hook into the ring on the bottom of the crucible, and pull back slightly. This will lock the crucible onto the lift handle. (See Fig. 2-10)

6. Move carefully to the sand molds that you have prepared earlier. Hold the crucible about 3-4" above the mold. Pull back on the hook to tip the crucible, and pour the liquid metal into the molds. NOTE: High leather boots or work shoes are the footwear for this job. NO TENNIS SHOES.

7. After the pour is completed, turn the crucible upside down, and rap the bottom with the hook to dislodge

most of the slag. Turn it right side up, and set it on the fire brick to cool.

8. Dump the furnace carefully. It is still radiating a lot of heat and can catch clothing on fire very quickly. Remove all of the charcoal. Remove the angle iron vent from the furnace. A 1/2" steel rod bent into the shape of a fireplace poker is best for this job. Set the furnace upright, replace the lid, and allow it to cool.

9. With a sprinkling can, quench still-burning charcoal (and place in another container). BE VERY CAREFUL OF THE STEAM GENERATED. Save this charcoal for a future run.

10. Allow the pigs to cool for about half an hour. Then carry the molds to the molding bench, and dump (shake out) the pigs and sand onto the unused molding sand still in the bench. Set the sieves and bottom boards aside. Using the mason's trowel, carry the hot pigs to the pouring area to cool.

Congratulations!! You have just made your first run.

Again, this process of melting dirty scrap and casting into pigs for later remelting improves the final castings.

Part Two: Making Ornamental Castings

Chapter 5
Flat Molds

THE PATTERN

The only remaining job is the construction of some molding boxes or flasks. These are simple to make and as your foundry grows, you will acquire a large variety of sizes and shapes of these boxes. The flask described here is one for general use.

I have found that a good quality 3-1/2'' wide furring strip, 8 feet long, that sells at lumber companies for under $1.00, will make a fine molding flask. A flask is actually a set of 4 pieces: (1) A top box called a ''cope,'' (2) A bottom box called a ''drag,'' (3) a board to cover the bottom of the drag, and (4) a board to cover the top of the cope. The cope has locating pins on each end. The drag has matching sockets so that the set can be assembled in exactly the same way each time.

Take the furring strip, and cut it into eight equal lengths (about 12'' long). Glue and nail them into two square boxes as shown in Fig. 5-1. The metal end brackets will give the flask more stability and should be included. Make sure that the boxes will set on each

33

other without rocking, and make sure the sides match. Do a good job building your flasks. Good flasks are needed to make good castings.

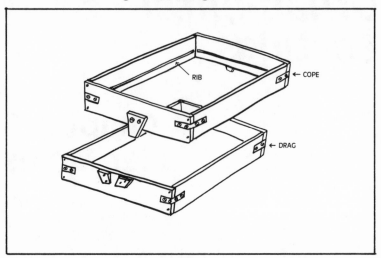

FIGURE 5-1 Molding box construction.

Into the cope add a 3/8" square strip of wood, called a "rib," all the way around the inside about 3/4" from the bottom. Use glue and small nails. This rib is necessary in the cope to keep the sand from falling out when the cope and drag are separated during the molding process. Cut two small pieces of 1/4" plywood into a tapered shape, and nail and glue them to each end of the cope as shown. You can offset one of the pegs a little so that it will be impossible to assemble the flask any other way. (See Fig. 5-1)

Set the cope on top of the drag, align the sides carefully and clamp gently with "C" clamps. Glue and nail to the drag sides two small pieces of 1/4" plywood that will mate with the cope pins. (See Fig. 5-2) After the glue has set, you may have to sand or file the cope pins a bit until both halves join tightly without play but separate easily. Finally, cut two squares of 1/4" or 3/8" thick plywood the same size as the cope and drag. Take the actual dimensions from

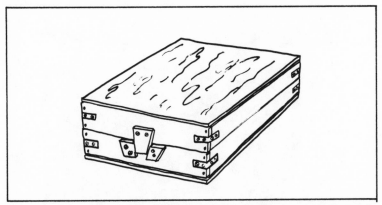

FIGURE 5-2 Cope and drag assembled with top and bottom boards in place.

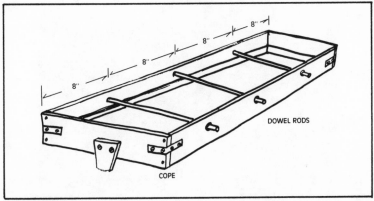

FIGURE 5-3 Cope with rib and extra cross ribs installed for more "hold" in the cope sand.

your flask. The boards must not only cover the edges of the flask but must also fit between the cope pins for certain types of molding exercises. (See Fig. 5-2) You now have a good, general-purpose molding flask.

You will soon discover a need for more and different size flasks. I have found that I use 3 or 4 flasks of this size on nearly every run. You can make 2 or 3 of these almost as easily as one. For flasks larger than 12" square, you will have to include cross ribs or gaggers in the cope to keep the sand from falling out. A simple method of doing this is to drill 3/8" holes in the sides

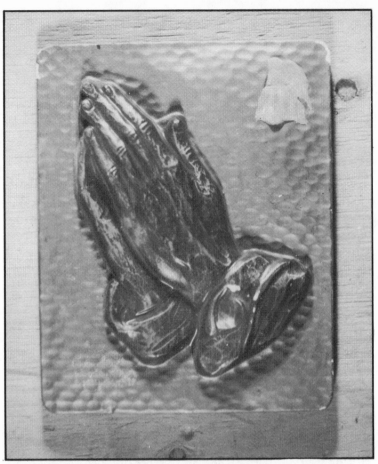

FIGURE 5-4 Plaque pattern on molding board.

of the cope and insert 3/8'' dowel rods before you ram up the cope. (See Fig. 5-3)

Flat molds are by far the simplest to sand-cast and are an excellent way to gain experience in molding and pouring. Almost anything can be used as a pattern for flat molds. For illustration, I'll use a praying hands plaster of paris wall plaque. This plaque has a parting line at the rear edge of the pattern. We'll discuss more about parting lines later in the next chapter.

Check your molding sand, using the squeeze test described in the last chapter to see if the sand is

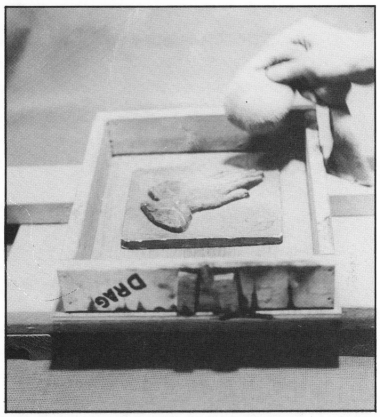

FIGURE 5-5 Drag placed on bottom board. Pattern is dusted with parting compound.

suitable for molding. Select a molding flask that will allow at least two inches of sand around the pattern. Take the bottom board and lay it down on the molding table. Place the plaque pattern face up on the board. (Fig. 5-4)

Next, place the drag (bottom of your molding flask) upside down on the bottom board. The cope pin sockets on the drag are facing the bottom board where the pattern is resting. Dust well with parting compound. (Fig. 5-5) Use an old wool sock, and fill it half full of parting compound. Tie the end, and shake it over the pattern. ALWAYS WEAR A FILTER MASK WHEN MOLDING BECAUSE THE PARTING COM-

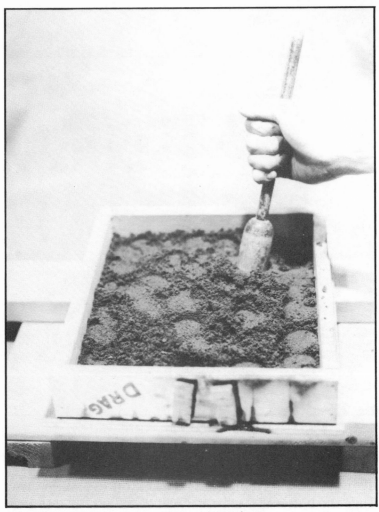

FIGURE 5-6 Tamp or "ram" sand into drag.

POUND AND THE SAND DUST CAN BE HARM-
FUL TO YOUR LUNGS.

Riddle about 2″ of sand into the drag, and ram
gently but firmly. (Fig. 5-6) Use the potato masher.
Continue to add sand, and ram until the drag is filled
slightly above the sides. Using the flat side of the pig
mold pattern made earlier, rap the sand smartly all
over. Take another bottom board, and shave the ex-

cess sand off until it is level with the sides of the drag. (Fig. 5-7)

Take the vent wire, and vent the pattern area 12 to 15 times. This is important! The wire is pushed through the sand all the way to the pattern. The resulting holes will allow steam to escape that is generated when molten metal meets damp sand. Without vents, you can expect to find holes and bubbles in your casting.

Sprinkle a little sand on the drag, and rub the bottom board back and forth on the drag. Grasp the drag and both boards firmly like a sandwich, and turn them over as a unit. (Fig. 5-8) This is called a "roll."

FIGURE 5-7 Shave off excess sand.

FIGURE 5-8 Rolling the drag, which is held firmly between two boards.

Remove the board. You should see the back of the pattern in the center of the sand. Using a trowel, remove any sand from the edges of the back of the pattern; the whole rear surface must be showing. Set the cope (top half of your molding flask) in place. Select a 1'' dowel or piece of pipe about 6-8'' long, and press it about 1/2'' down in the sand about 1'' from the pattern. Press another 1'' dowel or pipe into the sand about 1" from the pattern, but on the opposite side. The one pipe is the pouring sprue, and the other will be the vent. Dust the entire surface with parting compound. (Fig. 5-9)

Now carefully fill the cope with sand, about 2" at a time, ramming well all over each time and being careful to ram firmly in the corners. Level and smooth the sand at the top of the cope with a small trowel. Rap the vent and pouring sprue with the rapper (old end wrench), and remove them. Carve a pouring funnel in the sand by the pouring sprue. (Fig. 5-10) Care-

FIGURE 5-9 Cope placed on top of drag. One pipe is pouring "sprue," the other is a vent.

fully separate the cope from the drag, and set it on its side. Using a small vent wire, push it completely through the cope sand, starting on the pattern side, several times to vent the cope. (Fig. 5-11)

Now we turn our attention to the drag. Dampen the pattern edges with water using the syringe. Take a bent spoon and cut a 1/2" wide x 1/2" deep trench

FIGURE 5-10 Fill the cope with sand, ramming well every two inches. Level sand with trowel.

FIGURE 5-11 Use wire to vent cope sand.

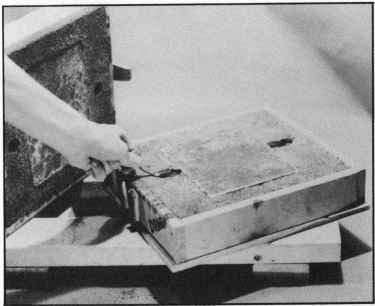

FIGURE 5-12 Use bent spoon to cut deep trench from pattern to sprue and from pattern to vent.

FIGURE 5-13 Remove plaque pattern from drag.

FIGURE 5-14 Replace cope on drag.

44

called a "gate" from the pattern to both the pouring sprue and the vent. (Fig. 5-12)

Using your fingers in the gates, move the pattern very slightly back and forth to loosen it from the sand. Lift it straight up and out of the mold. (Fig. 5-13) The remaining cavity should have a clear impression of the plaque and should be free from defects and loose sand. Loose sand can be removed with the moistened tip of a dowel, or blown out with an air bellows as described earlier. Small defects can be repaired gently using small trowels, old spoons and your fingers. But only small repairs.

Now for the tricky part. Gently pick up the cope, and invert it right side up. Move it over the drag, and lower it into place. Make sure that the locating pins mesh correctly. The mold is now ready to pour. (Fig. 5-14)

Carry the mold to the pouring area, and gently set it down on a level surface, preferably a dry sand pouring bed. Place a damp towel over the mold, and it will keep for a couple of hours. This will give you time to prepare two pig molds into which you can pour any leftover molten metal.

POURING THE FIRST CASTING

Set up the furnace as described earlier, and melt a couple of pigs of aluminum. When the charge has been stirred, skimmed, and is ready to pour, remove the damp covers from your molds. Make sure the area is clear of anything that would interfere with your movements—including onlookers.

Switch off the air blast, remove the top of the furnace, and carefully hook the crucible with the lift hook. Bring the crucible straight up out of the furnace. Slip the pouring handle into the ring on the bottom of the crucible, and carry it to the mold.

Pour the molten metal carefully but quickly into the funnel-shaped pouring sprue. (Fig. 5-15) Keep the

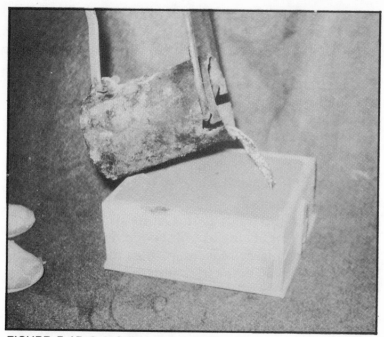

FIGURE 5-15 Carefully pour your molten metal into funnel-shaped pouring sprue.

sprue full of metal all during the pour, but stop as soon as the metal reaches the top and threatens to overflow the mold.

Pour any other molds that you have made. Pour the leftover metal into pig molds that have been made just for this purpose. Turn the empty crucible upside down in the sand bed, and rap it several times on the bottom to loosen the slag. At this point you should dump the furnace and clean it so that it will be ready for the next run.

Aluminum loses its strength at high temperatures. High strength machine castings should be allowed to cool completely before removing from the mold. Because ornamental castings do not require the strength of machine castings, they can be removed from the sand sooner. After 15 to 20 minutes, you can shake the casting out of the mold and view your creation.

FIGURE 5-16 Molding flask, pattern, mold and casting.

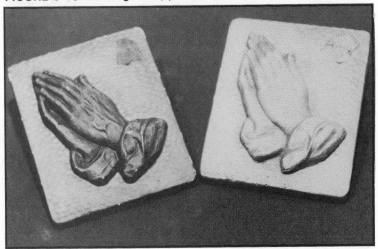

FIGURE 5-17 The original pattern and finished casting.

The casting should be a faithful reproduction of the original plaque. The surface will probably be rough due to the texture of the molding sand. (Fig. 5-16) When the plaque is cool, cut off the sprues and vents with a hacksaw, and smooth the cuts and casting edges with a file. The casting can then be buffed with a rotary brush attached to an electric motor to remove the sharp feel from its surface. USE A FULL FACE SHIELD FOR THIS OPERATION. At this point, you can spray on a coat or two of clear finish (lacquer or enamel) to prevent oxidation, and it will be ready to hang in some prominent place in your home. Fig. 5-17 shows the original pattern and the finished casting. This is a great way to convert fragile plaster figures into aluminum for outdoor use.

Chapter 6
The Sundial

PATTERN MAKING

The sundial is a good exercise in pattern-making and casting. The sundial pattern is simple and a good one to start with.

Obtain a 1 ft. square of 3/8'' plywood with a good finish on both sides. Use a compass to draw a 10'' circle on the wood, and cut out the pattern. If you use a hand-held electric jigsaw, you can set the blade to cut at a slight angle and make the draft in the pattern at the same time. If not, the draft can be sanded into the pattern afterwards. Draft is the slight bevel or angle that is in a pattern edge which facilitates its removal from the sand without damage to the sand mold. (Fig. 6-1)

After putting draft into the pattern, fill any holes in the edge surface with wood putty, and sand smooth. Now you're ready to attach the lettering to the pattern. Sundials are generally lettered for the part of the earth where they will be used. I copied mine from a 17th century French sundial, and it is fairly accurate for the latitude of Ohio.

I formed each letter from individually cut and beveled strips of pine about 1/8'' wide. You can do this too, but I must confess if I make another sundial pattern, I will go to the hardware store and buy a set of

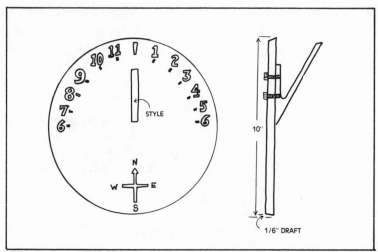

FIGURE 6-1 Sundial face and side view showing style attached.

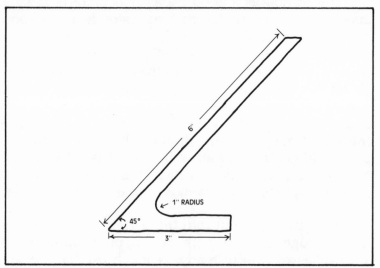

FIGURE 6-2 Side view of style pattern.

stick-on clock letters for $2.50 and save a couple of days' work. However you decide to letter the face of your pattern, fix the letters neatly in the approximate spacing shown in Fig. 6-1. Any inscriptions, compass points, etc., can be added as desired.

When the pattern face is complete, drill a 1/16'' hole part way through the center of the BACK of the pattern for a draw peg. Give the entire pattern a coat of clear finish and allow to dry overnight.

THE STYLE

In order to cast a shadow from the sun, the sundial must have a pointer or "style." Styles are often very ornate and complicated. For your first sundial, I suggest you make a very simple style.

The pattern is made from a scrap of 3/8" thick white pine board. It is a simple pattern to make and cast, requiring draft on one side only. Lay out the pattern on the pine board, using the dimensions shown in Fig. 6-2, and cut the pattern out with a jig or coping saw. Sand lightly, making sure that the pattern has draft to one side (like the sundial pattern), and give it a coat of clear finish.

MOLDING THE SUNDIAL AND STYLE

Use a molding box that will allow about 2" of sand around the pattern. The box used for the wall plaque may be large enough. If not, construct a molding box set measuring 12-14" on the inside. You will never have anough molding flasks, so building new ones will soon become routine.

When a suitable flask is at hand, prepare a false cope. This is made by laying the flask down on a bottom board, pegs down, and ramming up just as though it contained a pattern. Rub in a top board on the cope, and turn the cope over. Remove the bottom board, and place the sundial pattern face up in the center of the false cope. Place the drag half of the molding flask in place. Dust the sundial pattern and the exposed sand in the false cope with parting compound. Riddle in sand, and ram up the drag in the same way that you did with the wall plaque. Smooth off the excess sand and rub in a bottom board.

Turn the entire flask over. This is called a "roll." Remove the top board. Remove the false cope, and dump the sand back into the molding bench.

Replace the cope on the drag. Dust the surface with parting compound. Set two 1" sprues about an inch away from opposite edges of the pattern. Use 1" pipe for the sprues, and push them down in the drag sand about 1" to set them firmly. Ram up the cope. If you have trouble with the cope sand falling out when the cope is lifted to remove the pattern, add 2 or 3 dowel rods to help support this large expanse of sand. (See Fig. 5-3)

Smooth the top of the cope with a trowel, and remove the sprues. Remove the cope, and set it aside. Wet the sand around the pattern with the syringe, and screw a draw peg into the hole in the pattern back. Rap the draw peg gently several times in all directions to free the pattern from the sand. Lift (draw) the pattern straight up and out of the sand.

Inspect the impression left by the pattern in the molding sand. Look for any small defects. Some small defects can be repaired by gently using a small trowel or spoon tip, etc. Using the bellows, gently blow out any loose sand from the mold.

Using a bent spoon as a gate cutter, form gates (trenches in the sand) from the pattern area to the sprue. This prevents the pushing of loose sand into the pattern area. The gates should be about 1/2" deep and 1/2" wide.

At this point we will turn our attention to the cope.

Using a fine wire, vent the cope about 15 different places over the pattern area. Run the vent wire from the pattern side clear through the cope sand. Pull the vent wire completely through the cope. Carve a pouring funnel in either of the sprues, and blow off any loose sand with the bellows. Carefully turn the cope right side up, and move it over the drag. Lower the cope onto the drag, making sure that the guide pins fit together well. The mold is now considered to be "closed and ready to pour."

RAMMING UP THE STYLE

The style is rammed up in the same way as the sundial except that a single 1" sprue is used, and the pattern is gated on either side of the flat sides of the pattern. Since both castings require a little less than one quart of melt, you may as well ram up both and pour them at the same time. Vent the pattern area well with a fine vent wire.

When the patterns are ready to pour, fire up the furnace, melt and pour in the same manner as described in the previous chapter. Allow the casting to cool for about 20 minutes, and shake out. The sundial should be a faithful reproduction of the pattern. Cut off the sprues and gates. File the cut area smooth. The sundial and style can be buffed with a wire brush to remove the sand texture.

FIGURE 6-3 Completed sundial cast in bronze.

ASSEMBLY

File the bottom of the style until it rests squarely on the face of the sundial. Note the position in Fig. 6-3. Draw a mark around the style. Locate and drill two 5/32'' holes through the sundial in the area where the style will attach. Replace the style on the marked area, and clamp it in place. If you do not have a suitable clamp, have someone hold it firmly in place while you mark one of the hole locations on the style through a hole in the sundial.

Drill one 5/32'' hole in the style, and use a tap to cut 1/4''-20 threads in it. Enlarge the matching hole in the sundial to 1/4''. Bolt the style in place with a 1/4'' machine bolt. Drill the other hole in the style through the other 5/32'' hole in the sundial. Remove the style and tap the threads in the second hole. Enlarge the second hole in the sundial to 1/4'', and replace the style. Fasten with two 1/4'' bolts.

The sundials can now be given a coat of clear finish. Assembly holes for mounting the finished sundial can be drilled and tapped for whatever mounting place and mounts that you choose. The worst thing about this project is that all your friends will want you to make them one (for nothing, of course!). Fig. 6-3 shows a completed sundial cast in bronze.

Chapter 7
Solid Figurines

This chapter will deal with decorative figures produced using a variety of molds and cast as solid aluminum castings. The type of mold that you use will depend on the surface detail required and whether the figure was originally cast from a two-part pattern. The making of figurines is a bit easier than producing machine parts because we do not have to create a wood pattern and the specifications of size are very flexible. Half the fun of this type of casting work is finding the patterns because they are everywhere. All we have to do is learn to look for them.

THE PARTING LINE

Every pattern that is suitable for green sand molding or two-part plaster of paris molds will have a parting line. The parting line is an imaginary line that separates the pattern into two halves.

The parting line indicates which parts of the pattern will end up in which half of the sand mold. For instance, a ball would have a simple parting line that runs around the exact center of the ball like the equator runs around the center of the earth. It should be obvious that the parting line is the line formed where the two halves of the sand mold meet. If you look closely at childrens' plastic toys, you can find the parting line. This is a good way to learn how to find the parting line on any pattern.

Keep in mind that the pattern must always SLOPE AWAY FROM THE PARTING LINE in all directors. This is so that the pattern can be removed from the mold without damage to the sand impression.

FISHES, FROGS AND TIMID TURTLES

In the upcoming fountain chapter, we employ the use of several fish, a frog or two, and an occasional turtle to give the fountain a decorative and realistic motif. The patterns were quite easy to find, and even easier to cast.

The fish pattern was found at the local Goodwill Store and was a child's plastic bathtub toy. Molded plastic toys for small children provide a wealth of patterns for the sandmold techniques because they are usually formed in two-part injection molds and contain only large, exaggerated detail which is ideal for sand-casting.

Many of these toys are made of very rigid plastic and need no further treatment for sand-casting other than to ram up and pour. Some, however (usually your favorite ones) will be somewhat soft. These will deform in the sand mold and produce bad casting. You'll have to stiffen the pattern.

To do this, cut a hole in some area of the pattern, usually the bottom, and fill the entire pattern with plaster of paris and allow to harden for about 24 hours. The pattern can then be used as is and even tapped gently with a wooden mallet to bed the pattern in the drag.

CHANGING THE PATTERN

Sometimes a molded pattern will be difficult to remove from the sand mold due to a slight undercutting of the pattern. Such undercutting can be overcome in injection molds but not in fragile sand molds. This is not a great problem since it can usually be corrected by filling the area that is troublesome with auto body putty or plaster of paris. Smooth the filled area with sandpaper, and give it a couple of coats of lacquer or shellac before casting. There's no need to get upset if it does not come out right the first time. Just remelt it, and try again.

The fish that was selected for the fountain measures about 6'' long and is 3'' wide. This took about a pint of aluminum to cast, counting the metal required for the sprues and risers.

To cast a figure such as the fish, prepare a blank drag which is made in the same way that the drag was prepared for the flat mold but with the pattern being omitted. Using a small trowel, scoop some of the sand from the center of the drag to help in bedding the fish. To "bed" a pattern, firmly press the fish down to the parting line and refill any hollow areas around it and smooth the face of the drag.

Place the cope in place, dust well with parting compound, set two 1'' sprues, one on each side of the thickest part of the pattern, and ram up the cope. Lift the cope straight up, and set it to one side. Set the cope on its side rather than face down so as not to mar the sand or run the risk of the sand falling out.

Rap the pattern on all four sides, and remove it from the mold. Vent the cope and drag well with the vent wire, remove any loose sand. Form a pouring funnel in one of the sprues.

Close up the mold, and pour rapidly, keeping the sprue full all during the pour. Allow solid figures to cool at least 30-45 minutes before shaking out of the sand. If a casting is rough on one side, it is usually due to the bedding process. To cure this problem, roll the flask completely after ramming up the cope. Dump the original drag, and ram up another over the pattern which is still in the cope. Roll the flask again, separate the cope and drag, remove the pattern, etc.

Small to medium solid figures are a delight to cast in aluminum, pot metal or pewter. If problems such as shrink cavities or bubbles occur, review the chapter on preparing sand molds. Usually a slight change in sprue location or more generous use of vent wire will be enough to solve the problem.

When casting is cool, cut off the sprues, gates and risers. File the cut areas smooth. A wire brush mounted on an old electric motor is good for smoothing the sand texture and giving the casting a bright finish. REMEMBER TO WEAR SAFETY GLASSES OR A FULL-FACE SHIELD AT ALL TIMES, AND AVOID LOOSE CLOTHING AND SHORT SLEEVES. When the casting is finished, give it a couple of coats of lacquer to preserve the bright finish.

Chapter 8
Penny Banks -
Replica Castings

In the golden era of the cast iron foundry, one of the most enduring freebies for the foundry worker was the penny bank. Although originally made for stores or advertisements, the patterns were usually kept in the foundry. For Christmas or birthdays, even the poorest foundry worker could slip one or two of these molds in on a run. Then during a lunch break or after work, the castings could be finished and assembled. They made fine presents. The originals are becoming hard to find and quite expensive to collect.

The original brittle cast iron was easily broken and not all that many have survived over the years. If you have one of these banks or can borrow one, you can use the original as a pattern and make a replica. Aluminum is an excellent material for replicas. It is more durable than cast iron, and almost unbreakable.

The penny bank is a hollow, two-piece casting with a very irregular parting line, held together by one or two small bolts. Each half is cast separately. Ramming up each mold requires a double roll and extensive coping down.

The sand core that forms the hollow inside of each half should be on the drag of the mold. The cope is rammed up first, rolled and coped down to the parting line of the pattern. The drag is then rammed up on top of the cope, and the mold is rolled again. The mold is

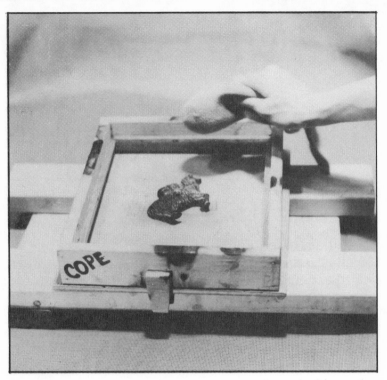

FIGURE 8-1 Half of bank pattern in cope, dusted with parting compound.

separated, and the pattern is removed. The cavity is gated and vented, closed, and then poured.

If the bank is a small one, both halves can be molded in the same flask with a central pouring sprue. For the first few times it is better to cast each half in a separate flask until you have mastered the coping down process.

The molding steps are as follows:

Remove the screw or screws from the bank, and separate the two halves. Select a flask that will allow at least 2″ of sand around the casting, both the sides and over the top. Set a bottom board on the molding bench, and set the cope half of the flask on it. Lay one half of the bank on the bottom board, and dust the pattern and bottom board with parting compound. (Fig. 8-1)

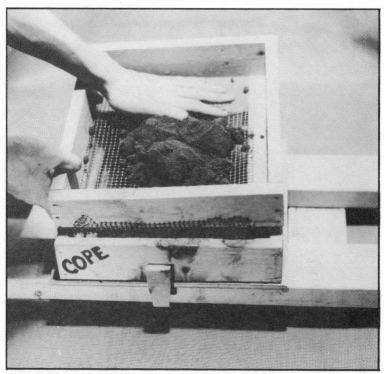

FIGURE 8-2 Riddle about 2″ of sand into cope.

Riddle about 2″ of sand into the cope. (Fig. 8-2) Set a 1″ pouring sprue about 2″ from the pattern. Banks are often gated between the legs on animal figures or near the heavier part of the casting. If you study the original, you can usually tell where it was gated. Riddle in more sand, and finish ramming up the cope. (Fig. 8-3) Rap, and remove the sprue. Rub in a top board. Grasp the whole mold like a sandwich (Fig. 8-4), and turn it over on the molding bench. This is the first of two rolls.

Remove the bottom board to reveal the sand and parts of the pattern showing through. (Fig. 8-5) Now comes the art of making penny bank sand molds: coping down. Using a small trowel, scrape away the sand to reveal the edges of the pattern. (Fig. 8-6) The legs, ears and tail and other parts must be exposed. Scrape

61

FIGURE 8-3 Ram up the cope. Pipe is 1″ pouring sprue.

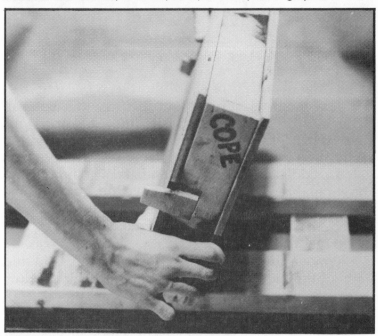

FIGURE 8-4 First roll of molding flask.

FIGURE 8-5 Half of bank pattern forming mold on bottom of cope.

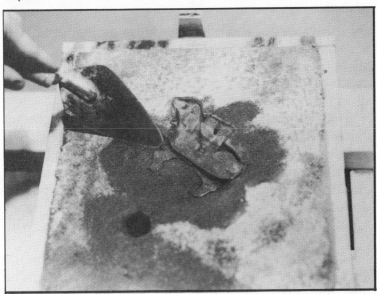

FIGURE 8-6 Use a small trowel to scrape the sand away from the edge of the pattern. The legs, ears and tail and other parts must be exposed.

63

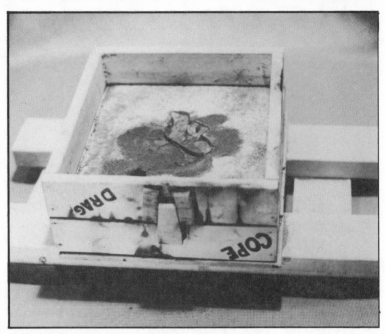

FIGURE 8-7 Dust the entire surface with parting compound. Place drag half of flask on the cope.

the sand away at a gentle angle from the pattern to the original sand level in the cope in such a way as to allow the cope and drag to separate easily. Smooth the sand surface with the trowel. Blow all loose and excess sand from the pattern and off the cope.

Dust the entire surface with parting compound. Place the drag half of the flask on the cope. (Fig. 8-7) Riddle in sand, and ram up the drag. Rub in a bottom board. Again grasp the entire mold like a sandwich and turn it over. (Fig. 8-8) This is the second roll of the double roll. If you find it hard to hold such a heavy mold, the mold can be clamped together with C-clamps and then rolled.

Remove the top board. Separate the cope from the drag, and set it aside. The pattern will be lying on the surface of the drag. (Fig. 8-9) Using a syringe, dampen the sand around the pattern edges, and gently rap the pattern in several directions to loosen it from the sand.

FIGURE 8-8 Make second roll of mold.

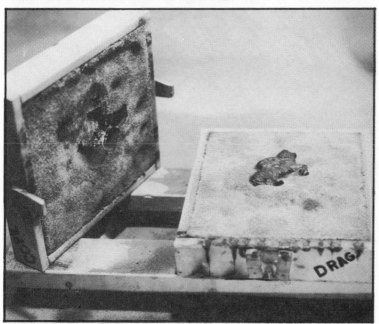

FIGURE 8-9 Separate the cope from the drag. Pattern is lying on surface of drag.

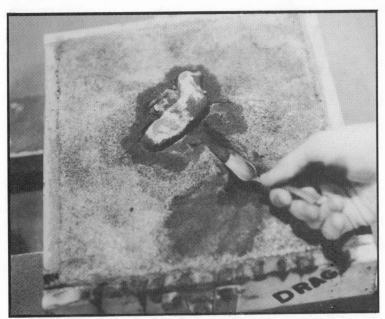

FIGURE 8-10 Dig a trench or gate from the pattern to the mark made by the pouring sprue.

FIGURE 8-11 Cut the pouring sprue again.

If you are using a cast iron pattern, a small magnet makes a great draw peg. Remove the pattern.

Using the moistened tip of a wood dowel, remove any loose sand that is in the pattern cavity. A small mound of sand or print will mark where the pouring sprue made contact with the drag. Using a gate cutter, dig a trench or gate from the pattern to this area of the drag. Dig a cup-shaped depression at the point where the sprue made its print. The gate should be about 1/2'' wide and 1/2'' deep. Smooth the gate and sprue basin with your fingers, and blow off any loose sand with the bellows. (Fig. 8-10) This completes the drag.

We will now turn our attention to the cope.

Using a thin wall 1'' steel pipe, cut the pouring sprue again. Steady the cope sand on the top side with the flat of your hand while gently cutting the sprue hole. You will be cutting through the original hole which has become filled with loose sand during the ramming of the drag. (Fig. 8-11)

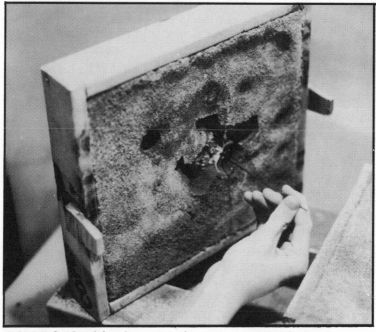

FIGURE 8-12 With wire, vent the cope in several places.

FIGURE 8-13 Carve a pouring funnel around the sprue opening in the cope.

Next select a 1/16'' vent wire, and vent the cope in several places, running the vent wire completely through the cope sand. (Fig. 8-12) Be certain to vent each leg and ear and to provide many vents in the body area. Blow off any loose sand with the bellows.

Move around to the front or top side of the cope, and using the trowel, carve a pouring funnel around the sprue opening. (Fig. 8-13)

Now very gently pick up the cope, invert it right side up. Move it over the drag, and lower it into place. Be certain that the locating pegs match. The mold is now ready to pour. (Fig. 8-14)

FIGURE 8-14 Replace cope on drag.

After the casting has cooled, cut off the gates and any vents. File the inside edge of each half until they fit together tightly without rocking. One half of the bank will have a hole for the bolt, and the other will have a pedestal cast on the inside of the bank into which the bolt will screw. A threaded hole must be created in this pedestal. I use a #7 drill followed by a 10-24 tap to cut the threads. If you are not familiar with the process of drilling and tapping holes, you should take the original bank or your replica casting to your local hardware store and purchase both a tap, the correct drill for that tap, and some stove bolts to match those in the original bank.

The process of creating tapped threads in the pedestal is as follows:

Drill the hole in the center of the pedestal as straight and deep as far as you can without drilling completely through. Drill a short way and measure. Drill a little more, and measure again. Take your time and get it right.

Once the hole is drilled, tighten the tap in the tap holder. If you do not have one, it would be wise to purchase one. Place a drop of oil on the tap. Start the tap into the hole just as you would a threaded bolt. Turn in the tap two full turns. Turn the tap backwards a half turn to clear out the chips being created. Then go in a full turn, back half a turn, in a full turn, back half a turn, again and again. Continue until the tap reaches the bottom of the hole. Back the tap out gently and completely. Dump the shavings out of the hole.

Take the other half of the casting, and try to insert the bolt into the hole in the side of the casting. Ream the hole if necessary. Countersink the hole so that the bolt head will be flush with the casting surface. Select a bolt that will pull the bank halves together with at least three turns.

Assemble your replica bank, and buff the entire surface with a wire brush to remove the sand texture. You can now paint the bank either in a life-like color or flat black like the original. Collecting and reproducing these old penny banks can be great fun, and they make excellent gifts for young and old.

Chapter 9
Plaster of Paris Molds

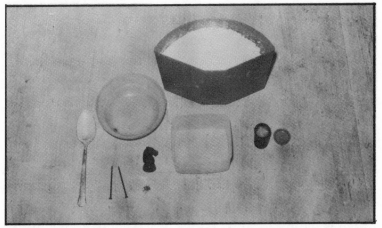

FIGURE 9-1 Items needed to make a plaster mold.

For small, solid figures with a lot of detail, a plaster mold may be superior to a sand mold. This is a simple process and requires only a few tools. (See Fig. 9-1)

You'll need a small amount of petroleum jelly such as Vaseline, a couple of 10-penny nails, a bowl and spoon, along with some plaster of paris.

You'll need a molding tray having sides which slope outward so that the hardened plaster of paris mold can be easily removed. Select a small square tray, approximately 3 times the thickness of the figure

71

FIGURE 9-2 Place the pattern in the center of the plaster and press it gently down to the parting line.

and at least 2″ larger than the figure on all sides. Coat the sides and bottom of the tray with a thin, smooth coating of Vaseline much the same as you would a cake pan before putting in a cake mix. Coat the figure with a smooth, thin coat of Vaseline. A small artist's brush is handy for this. Globs of Vaseline on the figure will ruin the fine detail.

Take two 10-penny nails and using a hacksaw, cut them into two pieces about 1/2″ back from the pointed end. These 1/2″ pieces will be your locating pins to insure that the mold lines ups (registers) correctly when you put it back together and pour the figure. Mix enough plaster of paris to fill the tray about half full. Mix the plaster of paris a little on the runny side. Pour the tray half full, and bump it gently on the floor or table several times to dislodge the air bubbles.

Place the pattern in the center of the plaster, and press it down gently to the parting line. Bounce the tray again, and check the pattern to be certain that it is

72

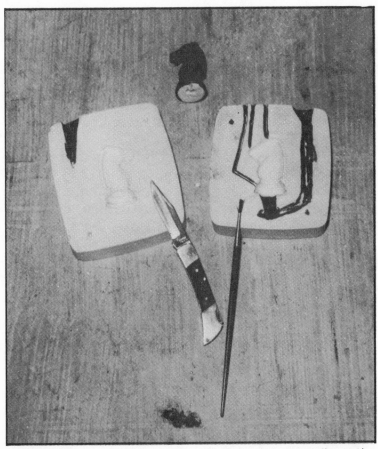

FIGURE 9-3 Using a sharp knife, cut a pouring sprue down the bottom half (side with the location pins) of the mold. Cut any vents or risers that seem necessary.

bedded up to the parting line, BUT NOT OVER IT. Place the cut nail ends into the wet plaster at any two corners that are diagonal from each other, with the pointed ends up. Press them down until only the points are showing. (Fig. 9-2) Set the mold on a level area, and allow it to set up for about 10 minutes.

When the plaster of paris is hard, gently remove any rough spots on the surface of the mold. Give the entire surface of the newly poured plaster mold and the pattern a light coat of Vaseline, being careful not to fill in any detail. Mix enough plaster of paris to finish filling the tray. Bounce the mold several times to help dislodge air bubbles from the plaster around the pattern. Allow the plaster to harden (about 30 minutes). Then remove the entire mold from the tray. Gently insert a thin-bladed knife at several places around the parting line of the mold until you can separate the mold. Remove the pattern, taking care not to damage the edges. (Fig. 9-3)

Using a sharp knife, cut a pouring sprue down the bottom half (side with the locating pins) of the mold. Cut any vents or risers that seem necessary. In Fig. 9-3 the sprue and vents have been darkened with a marker to illustrate their placements. Using a small brush and some plaster mixed to a very liquid consistency, repair any defects in the mold, such as air pockets and any damage that may have been done in removing the pattern. Do this sparingly. If the mold is badly damaged, it is better to make another than to try to salvage a bad mold. Give the sprue and vents a thin coat of plaster to make them smooth. Wet plaster surfaces can be smoothed even more by simply painting them with water.

Place the entire mold in an oven, and bake at 225° for 1 hour and then at 450° for 2 hours. As soon as the mold is cool enough to handle with gloves, fit the two halves together, and bed the entire mold in dry sand. A stout rubber band around the mold will keep it from separating during the pour. Pour the mold as soon as possible to keep it from absorbing water from the air.

These molds can be used several times. We were able to pour 14 chess pieces from a single mold. These molds are suitable for pewter, potmetal, or aluminum. NOTE: Pewter and potmetal will pour with a 3/8" or 1/2" pouring sprue, but aluminum must have a 3/4" or larger sprue.

I recommend pouring some of these type molds with lead or potmetal, using the gas ring and ladle, before trying aluminum or brass. If the mold spits back (metal flies back out of the sprue hole during the pouring process) or has other problems with potmetal, these will be 100% worse with the hotter metals.

The experience you have gained with sand-casting and the preparation of plaster of paris molds will be invaluable when making lost wax molds.

> **WARNING:** The process of baking the plaster mold in the oven forces out the water that is locked chemically into the crystalline structure of the plaster. If all the water is not removed, it might very well convert to steam when it comes in contact with molten metal. The result can be the dangerous boiling and sputtering of molten metal as the steam bubbles out. Explosions are even possible. The higher the temperature of the molten metal, the greater the danger of this boiling or spalling. For instance, just a few drops of molten cast iron dropped on concrete will cause an explosion. A drop of sweat into a ladle of cast iron is extremely dangerous. Be very careful in pouring these plaster molds! YOU HAVE BEEN WARNED.

Chapter 10
Lost Wax Molds

Lost wax molding has been both an art form and an industrial method for producing intricate castings for nearly as long as foundries have been in existence.

In the lost wax process, the figure is modeled in wax with fine detail. Unlike green sand molding, we are not concerned with the parting line, but we must be concerned about the wall thickness of the final wax figure. If an area of very thin wall runs into a thick-walled area, the thin area will rob metal from the thick part, causing a casting defect called a "shrink cavity." Rather than make any attempt to duplicate these books, let's simply make a lost wax casting and see how it works.

The wax figure can be anything you create from a low-melting temperature wax. Ordinary candle wax will work fine. You'll need a small alcohol lamp or other source of heat, and a small metal spatula. There are an infinite variety of artists' tools that are used in wax work that you can purchase, but most of these can be created as needed from old kitchen spoons, Ex-acto knives, bent clothes hanger wire, etc. The secret is that wax parts can be "welded" together with spare wax chips and shavings and a hot piece of wire or spatula blade.

For this exercise we will cast, using the lost wax process, a bible theme paperweight, "The lion will lay

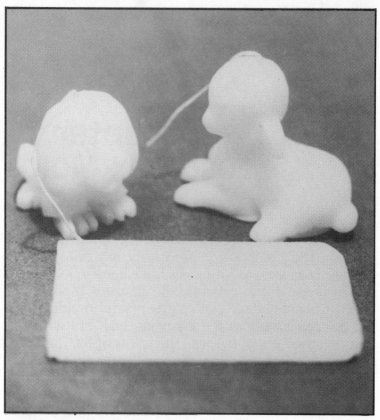

FIGURE 10-1 Candle figures used as patterns and square of paraffin candle wax used as base.

down with the lamb.''

Some of you are probably great sculpturers, but I am not. For this piece I went to my wife's friends who make candles and purchased a small lion and lamb figure. (Fig. 10-1) The base is a square of paraffin candle wax straight from a 1 pound box purchased at a local store. The bottom of each figure must be scraped smooth until they fit square and snug on their flat base. (Fig. 10-2) Be certain that each figure is sealed all the way around. Use wax shavings to fill gaps wherever necessary. Take your time with the step. Every small defect often seems to be magnified in the final casting.

When the figure is as complete and smooth as your patience will allow, you can start attaching the gates, sprues, vents and risers.

Here is where your experience in green sand molding will be invaluable. Think of each casting that you have tried and how it looked when it was shaken out of the sand with the gates, pouring sprues and vents still intact. This is what you must add in order to complete the lost wax figure. (Fig. 10-3)

In Fig. 10-3 we see one of the gates and sprues attached to the base and the parts of the second set ready to attach. I used 12″ white taper candles to make the pouring sprue and gate. Remember that the

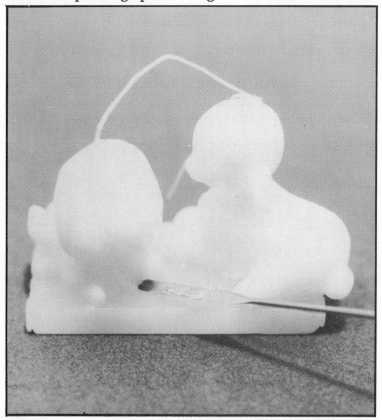

FIGURE 10-2 Use a heated metal spatula to fuse figures to base.

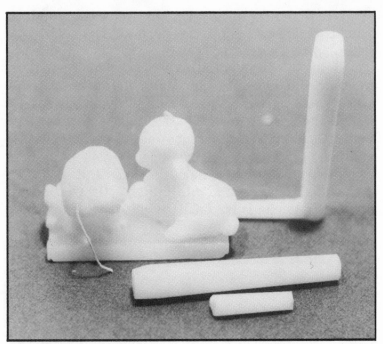

FIGURE 10-3 White table candles used to form pouring sprue and vent.

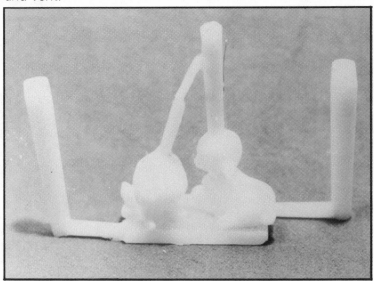

FIGURE 10-4 Short pieces of candle scrap attached to the head of each figure and joined at the top to form a common vent.

wicks from the candles must be removed from the plaster mold later on before pouring. For this figure we will create two equal pouring sprues on each side of the base. We'll pour into one and use the other as a vent and riser.

Since these figures are solid, they will be very hot in the head areas. The heads of both the lion and the lamb will require vents to eliminate gas bubble cavities and to act as risers to provide extra metal to compensate for shrinkage as the casting cools. I attached short pieces of scrap candles to the top of each head and joined them at the top in a common vent. (Fig. 10-4) Be certain that the risers and the sprues are of approximately equal height. At this point check your wax figure carefully, and smooth any remaining rough spots and fill any holes.

The next step is to coat the wax figure with a light coat of plaster of paris. Make the plaster very liquid so

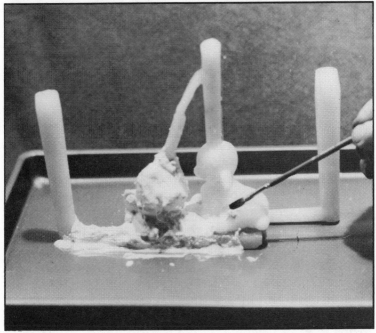

FIGURE 10-5 Coat the wax figures with a light coat of plaster of paris.

FIGURE 10-6 Let plaster-coated pattern dry for 24 hours.

that it will brush smoothly. (Fig. 10-5) Coat all parts of the figure including the sprues and risers. Build up the plaster to about 1/8" thick in 3-5 coats with a couple hours drying time between each coat. When the figure is completely coated including the bottom (Fig. 10-6) let it dry for 24 hours.

The next step takes very little equipment, but you will need total faith in the process. You must now destroy, by melting, the wax figure that you have put all this work into. When the wax drains out (is lost), a cavity which is our mold is left behind. This is the principle of "lost wax."

Since it is a good idea to melt the wax, bake the plaster mold, and pour the casting in the same day, a Saturday morning would be a good time to start. In order to melt the wax, make a simple rack or stand to hold the plaster mold upside down while it is heated in the oven. This will allow the wax to run out of the mold as it melts. Use a drip pan under the mold to catch the wax. I made a wood frame and loosely tied the plaster mold to it with some light gauge wire. (Fig. 10-7) The frame sets in an oven. Heat the oven to approximately 200° F., hot enough to melt the wax but not not enough to start a fire. (Since this is a definite fire hazard, it would be wise to have a fire extinguisher close by.) The plaster mold should be checked every 15 minutes.

When the wax stops dripping from the sprues and vents, tilt the rack several times in all directions to allow any wax that is still lying in small ledges and depressions in the mold to run out. Remove the pan and rack. The melted wax can be cast into cylinders, small squares for use in making the next figure, or can be used to make more vents, sprues and gates for later use. Using the two-part plaster molds described in

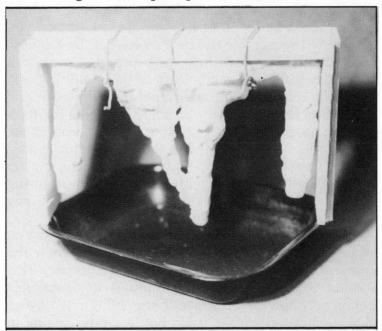

FIGURE 10-7 Prepare to melt the wax out of the plaster shell by suspending the mold from a wood frame over a drip pan.

Chapter 7, you can design molds to make any sort of wax shapes that you need for other projects. You may as well recycle the wax.

Carefully untie the plaster mold from the wood rack. Remove the loose candle wicks from the mold, and put it back in the oven. Handle the mold with the utmost care. It is very hot and very fragile. Use gloves. Set the mold on some aluminum foil on the oven rack, and bake at 450° F. for 1 hour.

While the mold is baking, you can select a green sand molding flask large enough to accommodate the plaster mold. You will also need enough clean dry sand to fill the space between the plaster mold and the molding flask that you have selected. At this point you can start to melt the metal that you have chosen to use for the final casting. It is a good idea to pour these plaster molds while they are still hot from the final baking and before they can absorb water from the air.

When the metal is nearly ready to pour and the plaster mold has been at 450° for at least 1 hour, you are ready to assemble the final mold. Set the molding flask on a bottom board near the furnace. Pour about 2'' of dry sand into the bottom of the molding flask. Carefully remove the plaster mold from the oven and place it on the sand in the molding flask. Fill the flask with dry sand, making sure that it flows around the plaster mold completely. Tap the flask lightly with a rapper to settle the sand tightly around the plaster mold. Fill the flask to within 1/4'' of any pouring sprues at the top, but being careful not to spill any sand down inside the mold itself.

Pour the metal into either sprue the same way you did for the flat mold in Chapter 2. Stop when the metal reaches the top of the opposite sprue. This casting will take at least 10 minutes to completely solidify so don't even move the molding flask. After that, you can move the flask, but leave the casting in the sand for at least 1 hour to cool and gain strength before shaking it out. Remember to use heat-resistant

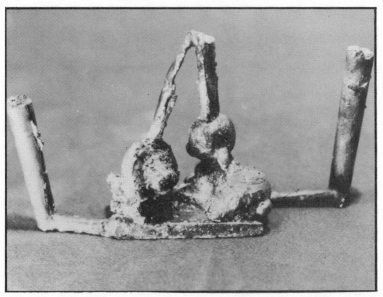

FIGURE 10-8 The metal casting after most of the plaster mold has been removed.

gloves even for shaking out, as this casting will remain hot for quite some time.

After an hour, shake out the casting. You'll find it still encased in the plaster mold. Using a rapper or light tack hammer, break away the plaster, and examine your casting. It should be perfect in every detail. Some of the plaster will be difficult to get off, but a short dunk in a bucket of water or a hosing off will quickly clean it up.

Take a minute to enjoy the satisfaction of having mastered a new skill. Let your chest swell with pride (but don't let your head swell). Having done this, critically examine your casting. Note any defects or mistakes, and take notes, mental or otherwise, on how to correct them in future castings. Fig. 10-8 shows the casting after cleaning off the majority of plaster. The gates, sprues and risers were left on to compare with the original. They will have to be removed and minor defects filled before polishing. Much work remains.

FIGURE 10-9 Chess piece cast in pewter, using the lost wax technique.

Remove the gates and risers with a hacksaw, and file the sawed areas smooth. File any rough edges around the base and on the figure itself. Lost wax castings should be very smooth, without the sand texture that one usually finds in green sand castings. After you have smoothed and buffed the rough places, give the casting a coat of clear finish, and it will be ready to take its place among your personal collection of art treasures.

Every casting is different and will require slightly different vents, riser placements, etc. Read all the books you can find, and don't be afraid to experiment. Use wax candle figures similar to the masterpiece you have in mind to gain experience before risking a hand-crafted original. Pour these trial castings with the same metal you intend to use for the final pour. You can always melt down the test casting and use it again and again. It's a lot of work but believe me, you'll be the only kid on your block who can do it. Fig. 10-9 shows two chess pieces cast in pewter, using the lost wax technique. Note the fine detail present in both figures.

Chapter 11
The Fountain

FIGURE 11-1 Decorative fountain, designed and cast by author and his wife.

One of the most rewarding projects for the backyard foundry is a working decorative fountain. The fountain will inevitably be of your own design, based on availability of space, patterns and your own taste. I will detail the basics of construction and show you what my wife Patty and I designed and built. (Fig. 11-1)

It's a lot of the fun meeting the challenge of working out the design and finding the patterns. It can become a giant scavenger hunt. We got great satisfaction creating a $2,000 piece of art for less than $200. The fountain project is divided into three basic areas: (1) the water system, (2) the base, (3) the figures and basin.

THE WATER SYSTEM

As is the case in any design, you must know exactly where you want to end up before you can start. So it is with the water system. After your ideas have been put down on paper and you know where the water outlets and the sump (where the water returns to for recirculation) will be, only then can you design the plumbing system. This system, or plumbing tree, must be completed and checked before anything else can be done. A typical plumbing tree would look like Fig. 11-2.

Make your plumbing tree from good quality material and standard fittings. Note that the outlets of different heights are on separate inlet pipes. (Fig. 11-2) This allows for balancing the water flow later on. If you use threaded pipe, assemble all joints with teflon plumbers tape. If you have not done any of this type of work, it might be best to have a plumber help you assemble the tree so that is is watertight and straight.

THE BASE

The base has a very simple function. It holds up the top of the fountain and hides the plumbing. It is also heavy enough to keep the whole fountain from falling over. For the base of our fountain, we used a plastic 5 gallon can. We cut the top off and cut a small hole in one side very near the bottom. Then we inserted the plumbing tree and blocked it to hold it upright and straight. Next, the entire 5 gallon can was

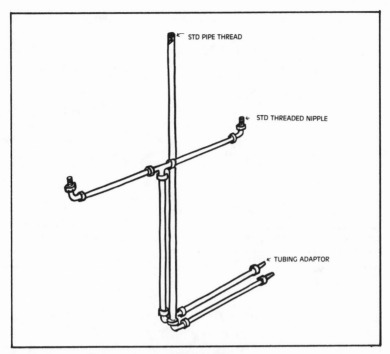

FIGURE 11-2 Fountain plumbing tree.

poured full of concrete. Several long nails were left sticking out of the concrete wherever necessary to help anchor concrete to be poured later. After 24 hours, the plastic pail form was cut away and discarded.

Next, forms were constructed for the arms. For this, 2 liter soft drink bottles were found to be just right. A 24" piece of 6" diameter aluminum pipe for the center section was set in place and well-secured with blocking and ties. The side forms were filled with cement, with about 6" of cement being poured inside the upright aluminum pipe. After 24 hours, the forms were removed.

The base, arms and the first 6" of the upright pipe were wrapped with a layer of chicken wire. A coating of 2 parts lime and 10 parts cement was troweled onto the the chicken wire as a white stucco. Using a small trowel, we attempted to reproduce the texture of tree

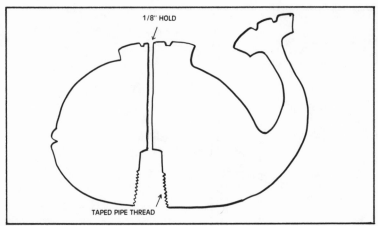

FIGURE 11-3 Section through center of fish figure, showing taped threads and spout hole locations.

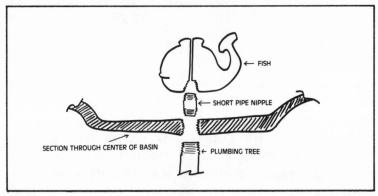

FIGURE 11-4 Exploded assembly diagram of fish and basin plumbing.

bark. The idea was to make the base and arms look like an old dead tree trunk. This was allowed to dry for several days before working on it any more. It is best to allow cement to cure slowly in a sheltered place out of the direct sun. This will prevent cracking caused by rapid drying.

THE FIGURES AND BASINS

My wife, Patty, found our patterns in a variety of second hand stores, junk shops and the local supermarket. The three basins were cast from a large glass relish plate. The fish was a child's plastic bathtub toy. The frogs, birds, and turtles were room deodorizers. The smaller basins were cast from a wooden nut dish shaped like a big oak leaf. And the dolphin was a plastic top for another fountain which we recast in aluminum.

Most of the basins required the construction of special size molding flasks, while the remainder were cast using the flasks already on hand.

The 3 figures which actually squirt water—two fish and the dolphin—required some extra work after casting. Each was prepared the same way. First, the base was drilled and tapped for a standard pipe thread. You can use whatever thread is on the pipe on your plumbing tree. next the 1/8" exit hole for the water stream was drilled from the top of the figure to the threaded hole in the base. (Fig. 11-3) In our design, the two side basin castings were also drilled and tapped with the same pipe thread as the figure to facilitate mounting and to make a watertight seal. (Fig. 11-4) Teflon plumbers tape was used on all joints.

The oak leaf basins were fastened to the upright pipe by designing a small right angle bracket and casting several in aluminum. The brackets were bolted to the basins and upright with 1/4" by 1/2" bolts. Holes were drilled and tapped in both the basins and upright pipe after a trial fitting. The other figures simply set in the basins or on the trunk and arms.

TESTING AND BALANCING

A water pump can be purchased at most places that sell commercial fountains. Ours cost approximately $45. You should obtain the following: a few feet of flexible Tygon or other plastic tubing, a couple of T or Y connectors, and 3 or 4 small metal clamps

(like small C-clamps or IV clamps).
Install the pump as follows:

1. Attach a short piece of tubing to the pump outlet.

2. Attach a Y connector to the tubing.

3. To one side of the Y connect a 3" piece of tubing and attach a screw clamp to the end of this tubing but do not tighten.

4. Now attach an 8" piece of tubing to the other side of the Y connector.

5. Using Y's or T's, connect this tubing to the inlets on the plumbing tree of the fountain. Attach a clamp to each inlet tube but do not tighten them

6. Add water to the fountain sump until the pump inlet is fully covered, and start the pump.

7. Slowly tighten the clamp on the pump by-pass tubing (step 3) until water streams are coming from all exit holes on the fountain.

8. Use the clamps on the inlet tubes (step 5) to adjust the height or pressure of the streams.

9. By adjusting the pump by-pass and the inlet clamps you can balance the water flow to your satisfaction.

Congratulations! Your fountain is now finished and ready to enjoy. If there is any one drawback in completing such a project, it is that you end up with 50 new ideas for use in building the next one. But that's the real beauty of ornamental casting. There's no end to it. Fortunately, the cost is slight. You'll need more time than anything else.